a little brown bat

The little brown bat is found in our land. It has wings, but it is not a bird. Bats are mammals, like people.

1

Little brown bats fly around
at night. They fly out of caves
and barns.

Little brown bats like to
eat insects. They swoop down
to get bugs.

Bats help us by eating
lots of bugs.

At night, a bat makes a high, clicking sound. This sound hits anything in its way.

Then the sound bounces back to the bat's ears. The sound tells the bat that something is there.

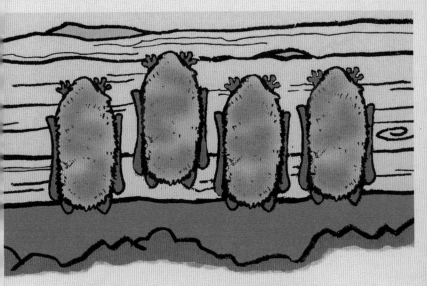

These four were flying
round and round. Now it is
time to sleep.

How do they sleep in the
daytime? They hang upside
down in dark places.

In winter little brown bats
sleep in caves. They do not
move about.

a bat house

Now people make houses
for bats. These bat boxes are
made from wood. They open
at the bottom. Bats fly in
and out of the boxes.

Now you see how bats
help people. In China people
think bats bring good luck.
Bats are good friends of
the earth.

8